SCRIPTURE IN THE AGE OF AI

BY DAN KEEN &
REV. JERRY CROSSLEY

HIGHER GROUND
BOOKS & MEDIA

Higher Ground Books & Media
Springfield, OH 45504
www.highergroundbooksandmedia.com

ISBN (Paperback): 978-1-971959-07-8

Printed in the United States of America 2026

SCRIPTURE IN THE AGE OF AI

BY DAN KEEN &
REV. JERRY CROSSLEY

Introduction

Artificial Intelligence (AI) isn't coming...it's here now, and this technology is moving forward rapidly. Advancements that were predicted to take years are coming to realization in only months.

You most likely used an AI program today: Did you do a "Google search"? Do you have "Alexa" or "Siri" in your home? Did you text someone on your cell phone, and as you were typing, the app suggested the next word before you typed it? Did you use the drive-thru window at your favorite fast-food restaurant? If so, you benefitted from artificial intelligence probably without even realizing it!

Society has always had to adapt to technological advancements, and there have always been people who resist it at first. Some people just don't like change. But for those who embrace it, they'll discover many benefits. Technology can make our lives easier, and give us more time to enjoy life.

Many technological advancements can be called "disruptive technology". When Alexander Graham Bell patented his technology for the telephone in 1867, the world was still wired to the telegraph. The telephone killed Western Union, who held the dominant share of telegraph lines across America. That meant telegraph operators were no longer needed.

Think about the time when horses and buggies were replaced by automobiles for transportation. (Poor Henry Hale Bliss was the first pedestrian killed in the U.S. by a car in New York in 1899! That's not how most of us would like to be remembered in history. But taking away his notoriety is Bridget Driscoll, who is the first pedestrian to ever be killed by a car. She was killed three years earlier in London.)

Consider how personal computers and word processing software killed the typewriter industry. More recent, Netflix's streaming service, which utilizes the Internet to deliver movies directly to people's homes, destroyed the Blockbuster movie rental chain. Apple's music streaming services killed Tower Records and most record stores. Elon Musk's Starlink service, delivering cell phone service from satellites, is currently threatening the cell tower industry.

AI is certainly a disruptive technology, and as with previous technologies, some people will lose their jobs as their particular jobs become obsolete. Even corporate jobs are being replaced by AI software in many companies. Huge companies like Amazon have replaced factory workers with robotic machines, scurrying around the warehouse, picking inventory from shelves and preparing packages for delivery to customers.

Yet, while some jobs become obsolete, others will be reshaped, and new jobs will be created. People must adapt and learn new skills. AI, combining with robotics, has freed people from many jobs that are hazardous, repetitive, mundane, or that nobody really wants to do.

Brace yourself for the new technology of artificial intelligence, which will benefit society as well as each of us personally. AI is transforming how we diagnose and treat illnesses. In education, AI's ability to adapt to individual needs can personalize a student's learning. AI is helping meteorologists make for accurate weather predictions. The financial markets use AI to assist in forecasting the stock market. In the workplace, repetitive tasks can be automated, freeing humans to do more creative and productive work. As the field of AI software integrates with humanoid robotics, we will see robots in human forms being used to care for the elderly and provide companions for those who are lonely, giving intelligent and stimulating conversation. In office environments, there will be hybrid teams, with people working along with AI programs. The Stitch Fix company doesn't involve shopping at brick-and-mortar stores, but instead, you fill out details about your size, budget, and clothing preferences on its website, and using AI alongside a human stylist, picks a handful of clothing items and mails them to you.

AI systems are increasingly making or influencing decisions about loans, hiring, criminal justice, healthcare, and content moderation. The values embedded in these systems – what they optimize for, whose interests they prioritize, how they handle trade-offs – reflect human choices. Christians and others who care about human dignity, justice, and the common good have perspectives that need to be brought to these design decisions. Therefore, we must learn about this new technology and be a voice in its implementation. We must have knowledge.

Knowledge is a Divine Gift

The Bible places remarkable emphasis on the pursuit of knowledge and wisdom, presenting learning not as optional, but as integral to a life well-lived and faithful to God. The Bible encourages practical knowledge and learning about the world. Solomon's wisdom included knowledge of plants, animals, and natural phenomena (1 Kings 4:29-34). The book of Proverbs contains practical wisdom about work, relationships, finances, and daily life. Throughout Proverbs especially, there's an almost urgent call to pursue wisdom and knowledge. "Get wisdom, get understanding; do not forget my words or turn away from them" (Proverbs 4:5). The text doesn't suggest passively waiting for knowledge to arrive, but actively seeking it out. Proverbs 18:15 reinforces this: "The heart of the discerning acquires knowledge, for the ears of the wise seek it out." Learning requires effort. The Bible thus presents a vision where learning is both a sacred duty and a lifelong journey – one that enriches our lives, strengthens our faith, and enables us to live more wisely.

Hosea 4:6 offers a sobering warning: “My people are destroyed from lack of knowledge,” indicating that ignorance has real consequences for how we live. Work is being redesigned around AI. Even if a job isn’t “AI-related,” it may be changed and reshaped. Workers today will need to learn how to work with new intelligent tools. Today’s young people would be wise to learn about AI and how it will affect job opportunities in the coming years, and to prepare now for employment in those fields.

In this book we’ll briefly explore how we can benefit from AI as a society, as citizens, and as Christians, and also consider some warnings. AI-literate citizens and Christians can act as a bridge between technology and human values. Join us on this journey, as we explore the exciting world of AI, its pros and cons, and its relationship to Christians today.

Chapter 1 - The Age of AI

Artificial Intelligence ("AI") refers to computer systems designed to perform tasks that normally require human intelligence. This software can recognize patterns, learn from data, make predictions or decisions based on instructions and data. It can create text, images, and even speech and music.

We interact with AI applications all the time, often without even realizing it. Perhaps you have recently done a "Google search" on your computer or phone. Have you visited a fast food restaurant? Taco Bell, White Castle, Wendy's and Taco John's use AI voice ordering systems. Several chains are also using AI camera systems to improve drive-thru operations.

Computer software programs have traditionally manipulated data by using a fixed set of instructions, and output that data in a form that is sorted to turn raw "data" into "information". For example, a "database" program might collect and store names, addresses and phone numbers of clients, and then be asked to sort that list alphabetically by name and address, or numerically by phone number.

AI software goes beyond that and can notice patterns across many examples, and even adjust how it responds. In other words, it can "learn" and adapt based on prior experience.

Such AI programs can be marvelous tools in countless situations. Imagine a program to control traffic lights based on predicting traffic patterns, recommending songs that a streaming platform is playing for you based on the type of music you have been listening to.

This "machine learning" ability allows the detection of patterns in large amounts of data in order to make future predictions or responses. For example, an AI program trained on millions of sentences learns how words tend to go together. When you are texting someone on your cell phone, you may notice that at times it will suggest the next word before you type it.

How does AI work? In the book “Simply Artificial Intelligence” (DK Penguin Random House ©2023, page 76), the process is briefly explained: “Artificial neural networks (ANNs) are machine learning modules based on algorithms [a sequence of instructions for accomplishing a specific a task]. Their structure is similar to that of the brain, consisting of interconnected nodes – artificial nodes – that are organized into multiple layers. The nodes within each layer receive, process, and send data to the next layer in the network, until an output or result is produced. Each node works like an individual microprocessor that can be reprogrammed to handle the data in a desired way. Using ‘training data,’ programmers can ‘teach’ an ANN to ‘learn’ how to give the expected results or outcomes.”

In the workplace, AI is changing how we approach productivity and creativity. By automating repetitive tasks, it frees people to engage in more creative and meaningful work. Rather than replacing workers, AI augments their capabilities, helping them analyze data, write, design, and make decisions more effectively. As AI software merges with robotics, humanoid machines, and autonomous systems, we’re standing at the threshold of a profound shift in how we live and work. The office and factory floor are becoming collaborative spaces, where humans work alongside intelligent physical systems. Autonomous delivery robots navigate warehouse aisles and office corridors, handling logistics, while workers focus on complex decision-making. In manufacturing, AI-powered robotic arms don’t just repeat programmed motions – they can adapt to variations, learn from human demonstrations, and coordinate with multiple systems in real-time. AI is creating entirely new industries and roles that didn't exist before, expanding economic opportunities.

Autonomous vehicles transform commutes into productive mobile offices. Humanoid robots in commercial spaces handle routine customer service, restocking, and maintenance. Autonomous vehicles are reshaping how families organize their days. Elderly parents can maintain independence longer with self-driving transportation. Parents send kids to activities without chauffeuring duties. Grocery shopping becomes a background task – you order, and autonomous delivery handles the rest.

The idea is to give humans more free time to pursue the good things in life. In factories, AI-enabled machines are freeing people from many jobs that are hazardous, repetitive, boring, or that nobody really wants to do. Manufacturers become more efficient with robotic machines that don't get tired, sick, or sleepy. South Korea has become the first nation in the world where one out of every 10 workers is a robot. New jobs will be created for people needed to develop, program, oversee and maintenance these systems.

AI is being used to write marketing copy, code software, act as customer service reps, assist financial analysts, paralegal assistants, and graphic designers.

Domestic life is gradually shifting toward ambient assistance. "Smart" building systems anticipate needs, adjusting environments and managing resources more efficiently (such as heating and lighting) with minimal human intervention. As robotics advances, smart home assistants are evolving into mobile helpers – robots that can navigate your home, handle laundry, prep ingredients for cooking, or tidy up spaces. These aren't the stuff of science fiction anymore; companies are already deploying early versions.

What remains constant is human centrality. "Physical AI,"

combining machines with software, excels at repetitive, dangerous, or physically demanding work, freeing people for more creative, strategic, and interpersonal activities. A barista's job (a trained coffee professional who specializes in preparing espresso-based drinks) might involve less grinding and steaming, and more conversation and craft. A nurse could spend less time on record keeping and transport, and more time providing compassionate care. Teachers could spend less time on paper work and preparing plans, and more time teaching students.

Such transition brings challenges – workforce adaptation, privacy concerns, ethical questions about autonomy and control. But the end result suggests a world where physical drudgery diminishes, where dangerous work becomes safer, and where human time increasingly focuses on what machines still cannot replicate: creativity, spending time with family and friends, and enjoying hobbies and activities.

AI can help society in many meaningful ways when it's designed and used responsibly. Across healthcare, education, environmental sustainability, and public services, AI is already making a positive impact, and its potential continues to grow.

In healthcare, AI is transforming how we diagnose and treat illnesses. By detecting patterns in medical images and data, AI helps doctors catch diseases earlier than ever before. Treatment plans can now be tailored to individuals based on their genetics, lifestyle, and medical history, moving us toward truly personalized care.

How about an AI therapy robot dog? Millions of people facing health adversities cannot safely or practically care for a live animal companion, but animal companionship has been shown to improve the quality of life for nursing home residents and elders with dementia. The company Tombot has developed a high-tech AI therapy robot dog named Jennie. Designed to sit on your lap, Jennie is a masterful combination of art, science, and engineering. Engaging with Jennie provides joy, comfort, companionship, and a sense of security in the midst of a destabilizing illness. Jennie's uniquely interactive autonomous behaviors mimic a real puppy. Such interaction is designed to alleviate the behavioral and psychological symptoms of dementia, including Sundowning Syndrome – a state of confusion common in dementia patients. Jennie can also help children manage pain, and autistic children speak more with a robot around. Jennie's not coming in the future...you could buy her right now at Tombot.com.

"Chatbots" are computer programs that simulate human conversation, either through text or voice, to provide automated assistance, information, or entertainment, using predefined rules for simple tasks or advanced AI processing for more complex, natural interactions. They can act as virtual assistants, answering frequently asked questions, handling customer service, guiding users, and performing tasks 24/7 across websites, apps, and devices. They work by following scripts and "decision trees" to provide specific answers. Using machine learning, they can "understand" intent, learn from interactions, and generate more dynamic, human-like responses.

Education is being revolutionized by AI's ability to adapt to individual needs. AI tutors can adjust to each student's pace, learning style, and specific gaps in understanding, making education more personalized and effective. For teachers, AI can automate time-consuming tasks like grading, lesson planning, and administrative work, freeing them to focus on what matters most: their students. Through translation, tutoring, and adaptive content, AI is making high-quality education accessible to lifelong learners everywhere.

AI is proving to be a powerful ally in protecting our planet. It improves climate modeling, helping us better predict climate change and extreme weather events. Smart grids and buildings powered by

AI reduce energy waste significantly. Conservation efforts benefit from AI's ability to track endangered species, monitor deforestation, and assess ocean health, giving us better tools to protect the natural world.

Communities are made safer when AI predicts natural disasters like floods, fires, and earthquakes, then optimizes emergency response to save lives. Financial fraud, cybercrime, and identity theft are detected faster through AI systems. Cities are becoming smarter as AI improves traffic flow, public transportation, and infrastructure maintenance, making urban life more livable.

AI has tremendous potential to create a more inclusive society. Speech-to-text, text-to-speech, and vision tools are supporting people with disabilities in unprecedented ways. Real-time translation is breaking down language barriers and connecting people across cultures. When used carefully, AI can even help identify unfair practices in hiring, lending, and policing, though this requires constant vigilance.

However, a word of warning: AI helps society only when it's guided by human values. Transparency and accountability must be built into AI systems from the start. Privacy and data protection cannot be afterthoughts. Fairness and bias mitigation require ongoing effort and attention. Most importantly, humans must remain "in the loop" for important decisions. AI is a tool – not a replacement for human wisdom, compassion, or responsibility.

Chapter 2 – How AI Can Be Used To Investigate The Bible

The advent of artificial intelligence presents both profound opportunities and challenging questions for religious communities. What happens when silicon meets scripture? Can machines help us understand ancient texts better?

As Christians, digging into the Word of God gives us great comfort, wisdom, and joy. AI can be used to investigate the Bible in ways that extend traditional biblical scholarship, but not replace it. Here are the main benefits, moving from technical to theological.

Since AI excels at pattern recognition across massive amounts of text, it can compare thousands of biblical manuscripts and identify variations in the text. This can help scholars assess which readings are most likely the original.

The Bible was written in the original languages of Hebrew, Aramaic, and Greek. AI can help interpreters avoid importing modern meanings into ancient words, analyzing syntax patterns unique to certain authors or periods. AI could map how a Hebrew or Greek word is used across the entire canon, and even identify semantic shifts over time. You can image the massive amounts of time such undertakings would be for a human!

It is well known that "Scripture frequently interprets Scripture". AI can reveal connections that are easy to miss when reading sequentially. A computer program could be written using artificial intelligence to identify Old Testament passages echoed in the New Testament. Themes such as covenant, sacrifice, exile, and kingdom could be traced across books.

An AI program need not be limited to one source, such as the

Bible. It can integrate biblical texts with external ancient sources to guard against interpreting biblical language (such as inscriptions and legal codes) belonging to a period other than that being portrayed. Place names and locations can be correlated with archeological data regarding events and rulers mentioned in Scripture.

Regarding literary style, there have been debates about composition of certain biblical writings. "Stylometry," the biblical analysis of variation in literary style between one writer and another, is a growing field, where AI is especially strong. AI can be used to examine the consistency within a book (Isaiah and Daniel, for example) without assuming theological conclusion.

AI can bring to light theological patterns without creating theology. It could show how justice, mercy, holiness, or wisdom are portrayed across Scripture, and analyze how divine attributes are described differently in various genres. This would help theologians see coherence and development across all books of the Bible.

Think how AI, with its ability to quickly analyze large amounts of data, would assist Bible readers in gaining understanding into translation decisions by comparing dozens of translations simultaneously. It could identify where translations diverge significantly, flag interpretive vs. literal translation choices, and highlight verses where theology heavily influences wording.

AI offers capabilities that biblical scholars could only dream of a generation ago. Machine learning can identify linguistic patterns, thematic connections, and structural relationships across thousands of pages of scripture in seconds. An AI program could trace how a single Hebrew word evolves in meaning across centuries, or map every instance of a particular metaphor through both testaments. This kind of comprehensive analysis would take human scholars years.

Imagine a seminary student using AI to quickly survey how different denominations interpret a passage, freeing the student to spend more time in contemplation and application. The AI handles the breadth; the human provides the depth. AI is a research assistant.

AI translation tools are making biblical texts accessible in languages that never had complete translations. For minority language communities, this democratizes access to scripture in unprecedented ways. Beyond simple translation, AI can help explain cultural context, idioms, and wordplay that gets lost between ancient Hebrew or Greek and modern languages.

AI tutor apps can help Christians by answering questions at appropriate depth, suggesting related passages, and adapting to individual learning styles. A teenager curious about justice can be guided to relevant prophetic texts; a philosopher pondering free-will can explore Paul's letters with contextual historical background.

In conclusion, AI is not a prophet. It is not a theologian. It is not a replacement for faith, tradition, or discernment. AI should be a partnership, not a replacement. Perhaps the future lies not in AI versus human understanding, but in thoughtful collaboration. Results should be reviewed carefully and cautiously. AI is an amazing tool, but interpretive authority should always remain human.

Here are a few AI-based websites that you can use right now in your study of Scripture:

GopselTruth.AI – This website can save hours of research for sermons, personal devotions, and Bible study.

BibleAI.com – Access an advanced Bible search engine here.

Tyhe.LY – Explore how AI can enhance your Bible study time.

ChatGPT.com – Here are some examples of a Bible-study "prompts" (the text a user enters) someone might use with ChatGPT AI website, written the way an actual student, pastor, or thoughtful reader might ask it:

• "Analyze Genesis 22 (the binding of Isaac) by explaining the historical and cultural context, key Hebrew words, and major theological themes. Compare how this passage is understood in Jewish and Christian traditions, and suggest two practical applications for modern faith and family life."

• "Break down Psalm 23 verse by verse, explaining imagery, original Hebrew terms, and how shepherding practices in ancient Israel inform the meaning."

• **For sermon or teaching prep:** "Help me outline a sermon on the Parable of the Good Samaritan, including background, main teaching points, and a modern everyday example."

• **For ethical reflection:** "How does Jesus' teaching in Matthew 5:38–48 challenge modern ideas of justice, retaliation, and forgiveness?"

• **For personal devotion:** "Guide me through a reflective Bible study on Philippians 4:6–7 with questions for prayer, trust, and daily anxiety."

Chapter 3 – The Intersection of Faith and Technology

Biblical understanding has never been purely intellectual; it's experiential and transformative. Sacred texts have always contained elements beyond pure logic: paradox, mystery, the ineffable. The burning bush, the incarnation, resurrection – these aren't puzzles to be solved, but mysteries to be contemplated. Can an AI, fundamentally a pattern-matching engine, appreciate what resists pattern? When scripture says, "Be still and know that I am God," the "knowing" isn't data processing.

For millennia, scriptural understanding emerged through community – rabbis debating in the Talmud, church councils wrestling with doctrine, small groups discussing meaning together. AI software cannot replace the friction of human disagreement, the wisdom earned through lived experiences, or the way interpretation deepens through relationship.

If understanding scripture involves spiritual discernment, grace, or the movement of the Holy Spirit, can an AI participate in that, or does it remain forever outside, cataloging words about transcendence without accessing transcendence itself?

It is important that we question who writes the software for these AI tools. If an AI app suggests an interpretation, what authority does that carry compared to a pastor, a scholar, or personal prayer? These questions will require wisdom to navigate.

If we consider that the Bible is primarily an information delivery system – a codex of facts and rules – then AI might eventually understand it better than we do. But if we believe that scripture is an invitation into a relationship with the divine God, a living word that speaks differently to each generation and each individual's heart, then AI remains a useful tool, but it cannot be the primary interpreter.

The technology is neutral; what matters is our wisdom in its use. An AI program can tell you every occurrence of "love" in the New Testament. But it cannot tell you what it feels like to be loved unconditionally, or how to love your enemy, or why love is greater than knowledge.

Perhaps that's exactly the insight we need: AI can enhance our study of scripture, but faith has always insisted that study isn't the endpoint. The goal is transformation, lived wisdom, encountering the sacred. Technology can light the path, but each person still has to walk it.

We must be aware that AI systems can perpetuate or amplify existing biases and inequities. This aligns with long-standing Christian emphasis on care for "the least of these." Many AI policy questions aren't purely technical – they're fundamentally about what kind of society we want. Should AI systems prioritize efficiency over employment? How do we balance innovation with safety? What relationships between humans and machines are healthy? These are philosophical and ethical questions, where religious and humanistic perspectives are as valid as technical ones.

If people with certain value systems are entirely absent from these conversations, the default values that get embedded may not reflect their communities' deepest commitments about human flourishing, meaning, relationships, and purpose. So as Christians, we must learn about this new technology, and be aware of its benefits and potential problems for individuals.

Chapter 4 - Biblical Boundaries for Artificial Intelligence

The rise of artificial intelligence confronts us with moral questions our ancestors never faced. Yet the fundamental issues at stake – responsibility, dignity, truth, community, justice, authority, and worship – are as ancient as Scripture itself. The challenge before us is not to force the Bible to address technologies it never mentions, but to apply timeless moral truths to new tools. When we examine AI through the lens of biblical teaching, seven clear moral boundaries emerge, each grounded in passages that speak directly to the ethical dangers these systems can pose.

Responsibility

Perhaps the most fundamental boundary involves the question of who decides and who answers for those decisions. Scripture consistently teaches that moral accountability cannot be delegated to anyone or anything else. In Deuteronomy 30:19, God commands His people to "choose life," placing the weight of decision squarely on human shoulders. Ezekiel 18:20 reinforces this principle with stark clarity: "The one who sins is the one who will die." The New Testament continues this theme in Romans 14:12, where Paul reminds believers that "each of us will give an account of ourselves to God."

These passages establish a principle that transcends any particular era: moral responsibility is inherently personal and non-transferable. Tools may assist our decisions, and advisors may inform our thinking, but ultimate accountability always rests with people. When AI systems are designed to make moral decisions on our behalf – determining who receives medical care, who gets hired, or who deserves parole – they cross a biblical line. Such systems may promise efficiency or objectivity, but they fundamentally violate the structure of moral agency that God established.

One problem with technology is that it progresses faster than governments can make laws and guidelines concerning them. For example, who should be responsible when an autonomous (driverless) taxicab runs over and kills a pedestrian? Should it be the back-seat paying customer, the owner of the car, the salesperson who sold it, the car's manufacturer, or perhaps the software company who wrote the AI computer code? It wasn't too long ago when we heard on the news that driverless cars were not stopping when school buses picked up and dropped off children. We also heard about autonomous cars being confused by the flashing lights of emergency vehicles, and blocking traffic at 4-way stops as they were unable to determine who should go first. Sure, software fixes can be made to avoid such scenarios from happening in the future, but serious tragedies may occur first before revealing these dangerous inadequacies. We must remember that when things go wrong, human beings should be ultimately responsible.

Humans Have Value

The second boundary concerns how we value human beings. Modern AI systems have the potential to dehumanize people, reducing them to data points, risk scores, or productivity metrics. This approach directly contradicts the biblical teaching that human worth is intrinsic and immeasurable. Genesis 1:26–27 declares that humans alone are created in God's image, bearing a dignity that cannot be quantified. Psalm 8:4–6 marvels at humanity's unique status, crowned with glory and honor. In Luke 12:6–7, Jesus teaches that while God cares for sparrows, people are valued incomparably more.

These truths establish that human dignity exists prior to and independent of any measurable qualities. We are not valuable because we are productive, intelligent, compliant, or efficient. We are valuable because we bear God's image. When AI systems treat people primarily as sources of data or obstacles to optimization, they ignore this foundational truth. The efficiency that AI promises must never come at the cost of treating people as mere numbers. Any system that overrides human dignity in pursuit of streamlined processes violates the biblical vision of what it means to be human.

The Danger in a System Designed to Deceive

Truth-telling forms the third biblical boundary for AI. The Bible treats deception as a fundamental moral violation, not merely a pragmatic problem. The ninth commandment forbids bearing false witness (Exodus 20:16). Proverbs 12:22 describes lying lips as an abomination to the Lord. Paul instructs believers to put off falsehood and speak truthfully to one another (Ephesians 4:25).

These commands apply with full force to AI systems, even when deception is subtle or well-intentioned. When algorithms are designed to manipulate behavior without users' knowledge, when systems generate content that misleads, or when AI interfaces create false impressions of understanding or care, they cross a moral line. Truth is not simply one value among many – it is foundational to trust, freedom, and love. The biblical standard is clear: any system designed to deceive, no matter how slightly or for what purported good, violates God's requirement for truthfulness.

My Robot Loves Me

The fourth boundary addresses one of AI's most seductive promises: the ability to simulate human connection. Yet, Scripture teaches that God designed growth and healing to happen through embodied relationship, not simulation. Genesis 2:18 reveals that isolation contradicts God's design: "It is not good for man to be alone." The author of Hebrews warns believers not to neglect meeting together (Hebrews 10:24–25). Paul describes his ministry to the Thessalonians as sharing not just the gospel but his very life (1 Thessalonians 2:7–8). These passages point to a profound truth: presence matters. The messiness, inconvenience, and inefficiency of human relationships are not bugs to be fixed, but features of how God works in the world. AI may support care – reminding the elderly to take medicine, connecting isolated people to resources, or providing information – but it must never be a substitute for the irreplaceable value of human presence. When AI companions are marketed as substitutes for friendship, when therapy chatbots replace counselors, or when algorithmic content replaces genuine community, we are trading what is real for what merely appears real.

Will AI become so sophisticated that it could lead people to believe the system has feelings and emotions? Will AI become so convincing that users might believe they even have souls?

Some "chatbot" users have become somewhat emotionally attached, as conversations progress deeper and become more complex and personal. Remember, as a conversation between a person and a chatbot continues, the chatbot becomes more knowledgeable about the user, and it can build upon the evolving conversation.

Discrimination

Justice forms the fifth boundary, and here the biblical witness is especially urgent. God's concern for the vulnerable runs throughout Scripture. Micah 6:8 summarizes His requirements: act justly, love mercy, and walk humbly with God. Proverbs 31:8–9 commands God's people to speak up for those who cannot speak for themselves. Isaiah pronounces woe on those who make unjust laws that oppress the poor (Isaiah 10:1–2). These passages demand that we examine how AI systems affect the powerless. When facial recognition systems misidentify people of color at higher rates, when hiring algorithms discriminate against women, when predictive policing concentrates enforcement in already over-policed neighborhoods, or when loan approval systems perpetuate historical inequities, the systems are not merely flawed – they are morally corrupt. God's justice includes how our systems treat the vulnerable. Any AI that harms the powerless, hides injustice behind the veneer of objectivity, or scales oppression with computational efficiency violates biblical principles.

How Much Authority Do We Give AI Results

The sixth boundary concerns the authority we grant to AI outputs. Scripture warns against treating any human creation as possessing ultimate wisdom. Proverbs 3:5 instructs us not to lean on our own understanding. Paul contrasts God's wisdom with human wisdom in 1 Corinthians 1:25. John commands believers to "test the spirits" rather than accept claims uncritically (1 John 4:1). These warnings apply directly to how we position AI in decision-making. When AI recommendations are treated as beyond question, when algorithmic outputs are implemented without human judgment, or when systems are designed to prevent human override, we have granted them an authority they cannot legitimately possess. No human-made system – no matter how sophisticated its training data or elegant its mathematics – possesses divine wisdom or moral authority. AI must always remain advisory, never authoritative. The moment we stop questioning its outputs is the moment we have crossed from using a tool to serving an idol.

Trust, But Verify

This brings us to the seventh and deepest boundary: idolatry and misplaced trust. The first commandment declares, "You shall have no other gods before me" (Exodus 20:3). The psalmist warns against trusting in chariots or horses – the advanced technology of the day (Psalm 20:7). Jeremiah pronounces a curse on those who trust in human power rather than in God (Jeremiah 17:5). Idolatry is not primarily about physical statues. It is about ultimate trust – where we look for security, meaning, and salvation. When AI becomes the source we turn to for answers about how to live, when we trust its predictions more than wisdom or community, or when we believe technology will solve problems that are fundamentally spiritual, we have made it a functional god. The allure is strong: AI promises knowledge without faith, certainty without trust, and solutions without transformation. But these are the same empty promises every idol has ever made.

We should not trust AI output without question. Although we can check facts to see if a result is true or erroneous, other types of "response" may not be so easy to verify. And, computers have great difficulty separating truth from untruth. They aren't "intelligent" as we think of intelligence in human terms, though they can simulate intelligent behavior in certain scenarios.

These seven boundaries suggest a framework for evaluating any AI system or application. We should ask: Does this honor human responsibility, or does it remove accountability? Does this protect human dignity, or does it reduce people to data? Does this promote truth, or does it manipulate and deceive? Does this strengthen community, or does it replace human connection? Does this advance justice, or does it amplify oppression? Does this respect proper authority, or does it claim false wisdom? Does this keep God at the center, or does it invite misplaced trust?

These questions flow directly from biblical passages that apply as fully to artificial intelligence as they did to the technologies and powers of ancient times. The specifics change, but the moral stakes remain constant.

Chapter 4 - Biblical Boundaries for Artificial Intelligence

The rise of artificial intelligence confronts us with moral questions our ancestors never faced. Yet the fundamental issues at stake – responsibility, dignity, truth, community, justice, authority, and worship – are as ancient as Scripture itself. The challenge before us is not to force the Bible to address technologies it never mentions, but to apply timeless moral truths to new tools. When we examine AI through the lens of biblical teaching, seven clear moral boundaries emerge, each grounded in passages that speak directly to the ethical dangers these systems can pose.

Responsibility

Perhaps the most fundamental boundary involves the question of who decides and who answers for those decisions. Scripture consistently teaches that moral accountability cannot be delegated to anyone or anything else. In Deuteronomy 30:19, God commands His people to "choose life," placing the weight of decision squarely on human shoulders. Ezekiel 18:20 reinforces this principle with stark clarity: "The one who sins is the one who will die." The New Testament continues this theme in Romans 14:12, where Paul reminds believers that "each of us will give an account of ourselves to God."

These passages establish a principle that transcends any particular era: moral responsibility is inherently personal and non-transferable. Tools may assist our decisions, and advisors may inform our thinking, but ultimate accountability always rests with people. When AI systems are designed to make moral decisions on our behalf – determining who receives medical care, who gets hired, or who deserves parole – they cross a biblical line. Such systems may promise efficiency or objectivity, but they fundamentally violate the structure of moral agency that God established.

One problem with technology is that it progresses faster than governments can make laws and guidelines concerning them. For example, who should be responsible when an autonomous (driverless) taxicab runs over and kills a pedestrian? Should it be the back-seat paying customer, the owner of the car, the salesperson who sold it, the car's manufacturer, or perhaps the software company who wrote the AI computer code? It wasn't too long ago when we heard on the news that driverless cars were not stopping when school buses picked up and dropped off children. We also heard about autonomous cars being confused by the flashing lights of emergency vehicles, and blocking traffic at 4-way stops as they were unable to determine who should go first. Sure, software fixes can be made to avoid such scenarios from happening in the future, but serious tragedies may occur first before revealing these dangerous inadequacies. We must remember that when things go wrong, human beings should be ultimately responsible.

Humans Have Value

The second boundary concerns how we value human beings. Modern AI systems have the potential to dehumanize people, reducing them to data points, risk scores, or productivity metrics. This approach directly contradicts the biblical teaching that human worth is intrinsic and immeasurable. Genesis 1:26–27 declares that humans alone are created in God's image, bearing a dignity that cannot be quantified. Psalm 8:4–6 marvels at humanity's unique status, crowned with glory and honor. In Luke 12:6–7, Jesus teaches that while God cares for sparrows, people are valued incomparably more.

These truths establish that human dignity exists prior to and independent of any measurable qualities. We are not valuable because we are productive, intelligent, compliant, or efficient. We are valuable because we bear God's image. When AI systems treat people primarily as sources of data or obstacles to optimization, they ignore this foundational truth. The efficiency that AI promises must never come at the cost of treating people as mere numbers. Any system that overrides human dignity in pursuit of streamlined processes violates the biblical vision of what it means to be human.

The Danger in a System Designed to Deceive

Truth-telling forms the third biblical boundary for AI. The Bible treats deception as a fundamental moral violation, not merely a pragmatic problem. The ninth commandment forbids bearing false witness (Exodus 20:16). Proverbs 12:22 describes lying lips as an abomination to the Lord. Paul instructs believers to put off falsehood and speak truthfully to one another (Ephesians 4:25).

These commands apply with full force to AI systems, even when deception is subtle or well-intentioned. When algorithms are designed to manipulate behavior without users' knowledge, when systems generate content that misleads, or when AI interfaces create false impressions of understanding or care, they cross a moral line. Truth is not simply one value among many – it is foundational to trust, freedom, and love. The biblical standard is clear: any system designed to deceive, no matter how slightly or for what purported good, violates God's requirement for truthfulness.

My Robot Loves Me

The fourth boundary addresses one of AI's most seductive promises: the ability to simulate human connection. Yet, Scripture teaches that God designed growth and healing to happen through embodied relationship, not simulation. Genesis 2:18 reveals that isolation contradicts God's design: "It is not good for man to be alone." The author of Hebrews warns believers not to neglect meeting together (Hebrews 10:24–25). Paul describes his ministry to the Thessalonians as sharing not just the gospel but his very life (1 Thessalonians 2:7–8). These passages point to a profound truth: presence matters. The messiness, inconvenience, and inefficiency of human relationships are not bugs to be fixed, but features of how God works in the world. AI may support care – reminding the elderly to take medicine, connecting isolated people to resources, or providing information – but it must never be a substitute for the irreplaceable value of human presence. When AI companions are marketed as substitutes for friendship, when therapy chatbots replace counselors, or when algorithmic content replaces genuine community, we are trading what is real for what merely appears real.

Will AI become so sophisticated that it could lead people to believe the system has feelings and emotions? Will AI become so convincing that users might believe they even have souls?

Some "chatbot" users have become somewhat emotionally attached, as conversations progress deeper and become more complex and personal. Remember, as a conversation between a person and a chatbot continues, the chatbot becomes more knowledgeable about the user, and it can build upon the evolving conversation.

Discrimination

Justice forms the fifth boundary, and here the biblical witness is especially urgent. God's concern for the vulnerable runs throughout Scripture. Micah 6:8 summarizes His requirements: act justly, love mercy, and walk humbly with God. Proverbs 31:8–9 commands God's people to speak up for those who cannot speak for themselves. Isaiah pronounces woe on those who make unjust laws that oppress the poor (Isaiah 10:1–2). These passages demand that we examine how AI systems affect the powerless. When facial recognition systems misidentify people of color at higher rates, when hiring algorithms discriminate against women, when predictive policing concentrates enforcement in already over-policed neighborhoods, or when loan approval systems perpetuate historical inequities, the systems are not merely flawed – they are morally corrupt. God's justice includes how our systems treat the vulnerable. Any AI that harms the powerless, hides injustice behind the veneer of objectivity, or scales oppression with computational efficiency violates biblical principles.

How Much Authority Do We Give AI Results

The sixth boundary concerns the authority we grant to AI outputs. Scripture warns against treating any human creation as possessing ultimate wisdom. Proverbs 3:5 instructs us not to lean on our own understanding. Paul contrasts God's wisdom with human wisdom in 1 Corinthians 1:25. John commands believers to "test the spirits" rather than accept claims uncritically (1 John 4:1). These warnings apply directly to how we position AI in decision-making. When AI recommendations are treated as beyond question, when algorithmic outputs are implemented without human judgment, or when systems are designed to prevent human override, we have granted them an authority they cannot legitimately possess. No human-made system – no matter how sophisticated its training data or elegant its mathematics – possesses divine wisdom or moral authority. AI must always remain advisory, never authoritative. The moment we stop questioning its outputs is the moment we have crossed from using a tool to serving an idol.

Trust, But Verify

This brings us to the seventh and deepest boundary: idolatry and misplaced trust. The first commandment declares, "You shall have no other gods before me" (Exodus 20:3). The psalmist warns against trusting in chariots or horses – the advanced technology of the day (Psalm 20:7). Jeremiah pronounces a curse on those who trust in human power rather than in God (Jeremiah 17:5). Idolatry is not primarily about physical statues. It is about ultimate trust – where we look for security, meaning, and salvation. When AI becomes the source we turn to for answers about how to live, when we trust its predictions more than wisdom or community, or when we believe technology will solve problems that are fundamentally spiritual, we have made it a functional god. The allure is strong: AI promises knowledge without faith, certainty without trust, and solutions without transformation. But these are the same empty promises every idol has ever made.

We should not trust AI output without question. Although we can check facts to see if a result is true or erroneous, other types of "response" may not be so easy to verify. And, computers have great difficulty separating truth from untruth. They aren't "intelligent" as we think of intelligence in human terms, though they can simulate intelligent behavior in certain scenarios.

These seven boundaries suggest a framework for evaluating any AI system or application. We should ask: Does this honor human responsibility, or does it remove accountability? Does this protect human dignity, or does it reduce people to data? Does this promote truth, or does it manipulate and deceive? Does this strengthen community, or does it replace human connection? Does this advance justice, or does it amplify oppression? Does this respect proper authority, or does it claim false wisdom? Does this keep God at the center, or does it invite misplaced trust?

These questions flow directly from biblical passages that apply as fully to artificial intelligence as they did to the technologies and powers of ancient times. The specifics change, but the moral stakes remain constant.

Chapter 5 - AI – Morally Neutral

Artificial intelligence stands before us as one of the most transformative technologies of our age, yet it arrives bearing a fundamental paradox. The technology itself possesses no moral character – it is neither virtuous nor wicked, neither just nor unjust. It is, in its essence, morally neutral. However, the ways we design, deploy, and depend upon AI are anything but neutral. Every choice we make about how AI interacts with human life carries profound moral weight. Understanding where AI crosses ethical boundaries, and why those boundaries matter, has become essential for anyone navigating our increasingly algorithmic world.

A dangerous line is crossed when humans delegate moral judgment to machines. This occurs when we allow algorithms to make consequential decisions about parole hearings, medical treatment, or even military targeting without meaningful human oversight. More insidiously, it happens when someone says "the AI decided" as a way to dodge accountability for outcomes that they would rather not own. Moral responsibility is fundamentally a human attribute. Scripture and ethical tradition consistently place moral agency on human shoulders, recognizing that conscience, discernment, and accountability cannot be outsourced to systems that lack understanding. When no one can clearly say, "I am responsible for this decision," we have already crossed into dangerous territory. The machine becomes a convenient shield against responsibility, and in that failure to fulfill a responsibility or duty, we lose something essential to our humanity.

AI crosses another critical line when it treats people primarily as data points, profiles, or statistical predictions rather than as beings with dignity and compassion. This dehumanization takes many forms. A programmer may build his biases into his software.

Predictive policing systems assume guilt based on patterns and correlations. Hiring algorithms quietly exclude candidates without explanation or appeal. Healthcare systems optimize for cost efficiency at the expense of compassion and individual care.

The ethical problem here runs deep. Human dignity is not statistical – it cannot be captured in data sets or reduced to probabilities. People possess inherent worth that transcends any algorithm's ability to measure or predict. When efficiency consistently trumps mercy, when the system's logic matters more than the person standing before it, we have allowed technology to fundamentally misunderstand what it means to be human. Any system that forgets this truth undermines the very foundation of human worth.

AI crosses into manipulation when it steers human behavior without informed consent. This happens through personalized misinformation campaigns, through algorithms deliberately designed to maximize addiction, outrage, or fear, and through "deepfakes" that distort truth and destroy trust. The distinguishing feature of manipulation is that it bypasses free will – the user doesn't realize they are being influenced, or the extent to which their choices are being shaped by invisible forces. ("Deepfakes" are realistic, AI-generated or manipulated videos, images, or audio that can make people appear to say or do things they never did, using deep learning to swap faces, synthesize speech, or alter expressions, creating convincing but fake media for purposes ranging from entertainment to serious disinformation, fraud, and reputation damage.)

Truthfulness forms the foundation of trust. Deception, even when justified by claims of positive outcomes, corrupts moral integrity. When technology operates to influence rather than inform, to control rather than empower, it violates the human capacity for genuine choice. This is not merely a technical problem, but a fundamental assault on human freedom and dignity.

Technology crosses a relational line when it replaces human presence, care, and guidance with artificial companionship. This might look like AI replacing human caregivers for the elderly or emotionally vulnerable, children forming their primary emotional bonds with AI companions rather than people, or individuals turning to artificial "spiritual guidance" that bypasses the accountability and formation that comes through genuine community.

Humans are fundamentally relational beings, formed and shaped through real connection with other people. Love, empathy, moral formation, and spiritual growth all require authentic relationships. While AI can supplement certain tasks, it cannot replace the irreplaceable dimensions of human presence. When convenience replaces commitment, we lose the very relationships that make us fully human.

AI does not create injustice from nothing, but it possesses a terrible power to amplify and scale the biases, inequalities, and exploitation already present in human systems. This line is crossed through surveillance tools that disproportionately target the poor and marginalized, through AI trained on biased data without meaningful correction, and through labor automation that enriches a small elite while discarding vast numbers of workers without consideration of their dignity or welfare.

Technology multiplies the moral impact of human flaws. An unjust decision made once by a person affects one situation, but an unjust algorithm deployed at scale can perpetuate harm thousands or millions of times over. Justice requires active, intentional protection of people who may be vulnerable. When those harmed by AI systems have no voice, no recourse, and no power to challenge the systems that affect them, we have allowed technology to become an instrument of oppression rather than flourishing.

Potential danger comes when we treat AI's outputs as unquestionable truth. This can happen when people regard AI medical or legal advice as definitive, when they use AI interpretations of Scripture or ethics without discernment and critical engagement, or when human judgment is simply replaced with uncritical acceptance of "what the model says." The fundamental error here is confusing knowledge with wisdom. AI can process vast amounts of information, identify patterns, and generate sophisticated responses, but it does not possess wisdom, conscience, or genuine understanding. It cannot weigh competing values with moral discernment. When disagreement with AI is dismissed as ignorance, when we grant algorithms an authority they neither deserve nor can responsibly bear, we have abandoned our responsibility to think critically.

The biblical and ethical tradition reminds us that tools are meant to serve humanity; humanity is not meant to serve tools. When our fear of losing AI outweighs our fear of losing truth, freedom, or compassion, we have allowed the created thing to usurp the place that belongs to our highest values and deepest commitments. This is the most subtle and dangerous crossing of all, because it reorients our entire moral universe around the wrong center.

Navigating the moral landscape of AI requires constant vigilance and careful discernment. Before embracing any AI system, we must ask several essential questions. Who is accountable when things go wrong? Who benefits from this technology, and who bears its costs? Does this system strengthen or weaken authentic human relationships? Does it honor truth and informed consent, or does it manipulate and deceive? Does it respect human dignity, or does it sacrifice the person on the altar of efficiency?

If these questions cannot be answered clearly and honestly, the moral line is already being approached – or has already been crossed. The challenge is not to reject AI wholesale, but to deploy it with wisdom, restraint, and constant moral attention.

The real moral danger is not that artificial intelligence becomes too intelligent, but that we humans become less responsible. We must not allow the sophistication of our tools to diminish the seriousness of our moral obligations. The moral lines we've explored – responsibility, dignity, consent, relationship, justice, authority, and worship – are not arbitrary restrictions, but guardrails that protect what is most precious about human existence. They remind us that technology, for all its power, remains a tool in human hands. The question is not what AI can do, but what we will choose to do with it. In that choice lies our moral responsibility, and in that responsibility lies our humanity.

From a faith perspective, AI is best understood as a tool entrusted to humanity, not a source of truth or moral authority. Wisdom belongs to God; technology belongs to human stewardship. AI can serve faith well by helping with scripture study, comparing translations and historical context to support deeper understanding without replacing prayer or discernment. It can assist churches in pastoral care, helping identify needs for meals, visits, and prayer requests so that people are cared for more effectively. Translation tools enable global ministry, sharing teaching and Scripture across languages. The key guardrail is clear: AI should assist reflection, not replace conscience, prayer, or community accountability.

Enjoy AI's Benefits

Healthy uses of AI applications in family settings include automating schedules, reminders, and budgeting so families have more time together. It can help children with homework by providing explanations (not answers), encouraging curiosity and confidence. Safety tools can monitor online risks while teaching wisdom and responsibility rather than fear. However, AI should never replace parental presence, discipline, or emotional connection. A simple rule guides this balance: if AI saves time, give that time back to people!

These principles come to life in ordinary situations. A parent uses AI to understand a medical diagnosis better before talking with their child's doctor. A teacher uses AI to adapt lessons for students who struggle, ensuring no one is left behind. A church leader uses AI to organize an outreach program, making sure the lonely and elderly aren't forgotten. A worker uses AI to reduce burnout by handling routine tasks, allowing him to focus on more meaningful contributions. In each case, AI supports human responsibility, rather than replacing it.

So, here's the "Big Picture". At its best, AI helps society by amplifying human potential, reducing inequality, and solving problems too complex for humans alone, all while keeping people firmly in charge. Across faith, ethics, and family life, one guiding question keeps AI in its proper place: Does this use of AI help people love God and love others more faithfully? When the answer is yes, AI can be a blessing. When the answer is no, restraint becomes wisdom.

Chapter 6 – Use AI Today to Benefit You Personally

With smart cell phones and computers, AI apps can assist you in quite practical, every day ways. The applications shown below are accessible from any web browser. In chapter 2, we listed several websites useful for Bible study, teaching or for writing sermons. Consider the following possibilities using the popular AI website **ChatGPT.com**,

Getting things done faster – You can use ChatGPT to draft emails, summarize long documents, help brainstorm ideas, or work through problems. It's like having a thoughtful assistant available anytime to bounce ideas off or handle tedious tasks.

Learning and understanding – Whether you're trying to learn a new skill, understand a complex topic, or get a different perspective on something, AI can explain things in different ways until they click, or help you work through problems step-by-step.

Creative work – AI can help with writing (from resumes to creative stories), coding, analyzing data, creating structured plans, or even just helping you think through creative challenges.

Personal projects – Need to plan a trip, organize a project, create a workout routine, or figure out how to approach a "do-it-yourself" task? AI can help structure your thinking and provide relevant information.

Daily problem-solving – From "how do I word this tricky message" to "what should I consider when making this decision," AI can be a sounding board for the small and large questions that come up in daily life.

The key is that you can interact conversationally, in our everyday speech (“natural language”). You don’t need to know special commands or formats. You can ask follow-up questions, ask for different approaches, or say “that’s not quite what I meant” and refine from there.

Information Generating Apps

ChatGPT.com

ChatGPT is an AI computer application used to find information, much as you would doing a “Google search”. ChapGPT, however, is built on the Generative Pre-trained Transformer model (thus the “GPT”), which understands “natural language”. This enables it to actually carry on “conversations,” since it “understands” the context of what is typed in. Unlike a simple one-sentence Google search line, ChatGPT keeps track of context as the conversation between the user and the app progresses. Furthermore, ChatGPT can answer more complex questions, and even provide insights. It can help you brainstorm, write emails or compose marketing copy. Have fun with it...it’s free to use!

Clickup.com

Clickup is an all-in-one productivity and project management app that helps teams plan, track, and manage work in a unified space, combining tasks, documents, goals, and chats. It’s free.

Otter.AI

Otter is an AI-powered app that automatically transcribes your conversations into searchable text. It’s a great note-taker for meetings, lectures, interviews, and the like. Otter is free.

Claude.AI

Claude AI is a next-generation AI assistant built by Anthropic, designed for natural conversations, content creation, coding, and complex reasoning. It is distinguished by its "Constitutional AI" framework focused on safety, ethics, and helpfulness, allowing it to handle tasks like summarizing large documents, analyzing images, and generating code. Claude acts as a smart assistant for brainstorming, drafting, editing, and answering questions. It can understand and process text, images, and even audio inputs, generating text, code, diagrams, and more. Claude is guided by a set of ethical principles to ensure safe, helpful, and unbiased responses, setting it apart from other models. It can analyze extensive documents, allowing it to understand and answer questions about entire books or large reports. For coding assistance, Claude excels at writing, debugging, and explaining code across various languages. It's free to use.

Gemini.google.com – Google's Gemini is a multimodal generating AI model, acting as an intelligent assistant for daily use at home or business. It understands and generates text, computer code, images, audio and video. It's free.

AI Image Apps

"AI image generators" are powerful software tools that use artificial intelligence to create unique images simply by the user entering a text description of what is wanted. These text descriptions are called "prompts". An example of a prompt would be: "Create a photo-realistic scene of a mouse who comes across a mouse trap that holds a big chunk of cheese, but the trap is ready to be sprung. In the background, show a stock market graph." This picture was used by a writer to emphasis options trading should be done by "nibbling like a mouse," that is, don't gamble on making big, risky trades, but rather place many small trades with less risk, as their profits will add up significantly over time.

Do a Google search for "AI image generators" and you'll find many incredible, excellent image generators. Some will create videos and even make a person or character speak words you want by entering text ("text to speech"). Such creation software was previously only available to Hollywood's motion picture industry. Today, you can do the same thing at home.

Many AI image generating apps require a fee, but most have free trials. Some sites go beyond simply creating a picture or image, allowing you to upload a photograph you may have and bring it to life with motion, creating a short video. It is even possible with some apps to enter text that you want an AI-generated character to speak. A great example of this can be seen on YouTube. Business owner Julia McCoy uses AI image generation to create an "avatar" or "clone" of herself. She creates videos on the subject of AI. By using her avatar, she can produce videos without having to set up cameras and props each time she makes a video. You'll be fascinated and gain great insight into AI by watching some of her videos. Go to YouTube and search for "Julia McCoy AI". It's hard to believe you're not watching an actual person!

Grok.com

Grok is an AI chatbot, known for its real-time knowledge from the social platform X (formerly Twitter) founded by Elon Must, and a witty, rebellious personality inspired by the book "The Hitchhiker's Guide to the Galaxy". As an AI assistant, Grok can help answering questions, brainstorming, coding, and reasoning. Grok features real-time access to information from the X platform, allowing it to answer current events. Grok offers text/voice interaction, image generation (with recent restrictions), document summarization, and coding assistance. The name comes from the verb "grok," meaning to understand intuitively and deeply, coined by Robert A. Heinlein. It's free.

Pixelcut.AI – If you have a photo of someone, but you don't like the background, Pixelcut can remove it! Simply upload a JPG of your picture. It's free.

Music and Voice Apps

Remusic.AI – Want to make your own karaoke songs? "Vocal remover" websites let you upload a song and remove the vocal. Remusic has an advanced separation feature that isolates vocal, bass, drums, guitar, and piano. It also includes a music analyzer to uncover key insights into a song's rhythm, melody, tempo in beats per minute (BPM) and more. In addition, its AI lyric generator can inspire lyrics for your next song.

Tunee.AI – Create "voice overs" with royalty-free vocals or train your own voices. Create copyright-free cover vocals, and much more.

Audimee.com – Audimee.com is an AI-powered vocal conversion platform for musicians and music creators, allowing users to transform vocals using royalty-free AI singers, clone the users own voice, generate harmonies, and isolate stems for music production, enabling high-quality vocal creation and enhancement with features like pitch correction and genre variety. It serves as a tool for making cover songs, producing music, creating backing vocals, and offering creative control over vocal tracks with professional-sounding results. Features include AI voice conversion to turn any vocal recording into a different voice, including male, female, and various styles like pop, rap, or dance. You can generate backing vocals, harmonies, and vocal stacks from a single vocal sample. It also includes vocal isolation and editing to separate vocals from music and use tools to adjust pitch and enhance clarity.

Personal Finance

Google.com/finance/beta – For those who do their own stock market investing and trading, this website focuses on business and financial news, providing real-time market quotes, company data and stock analysis.

PortfolioPilot.com – This site offers complete financial advice for self-directed investors, with no commissions and no conflicts of interest. Sophisticated financial models empower individuals to "do better themselves". The basic plan is free.

Bankrate.com – Bankrate has lots of expert tools, guidance, and descriptions of popular sites and their fees for further exploration.

Regarding the Job Market

For those seeking employment or who will still be in the workforce in the next few years, preparing for it now is essential. That topic goes beyond the scope of this book, but you will need to be prepared for the jobs of the future. We recommend you watch a 20-minute video on YouTube by Jensen Huang, the co-founder, President and CEO of Nvidia, the leading designer of AI semiconductors. Watch it at:

www.youtube.com/watch?v=jkDu_8wWcRY

In part, here is what he foresees:

"Nobody is ready for what AI can do in 2026, and I don't mean "ready" like you haven't updated your phone or you haven't the newest chatbot. I mean ready in the way people weren't ready for electricity, weren't ready for the Internet, weren't ready for the smart phone. Ready is the wrong word, because what's coming isn't a feature you enable. It's a platform shift. It's the kind of shift where the rules you grew up with believing – about school, careers, business, money and power, and even language –suddenly stop working the way you expect."

"The biggest mistake most people will make in 2026 is trying to use AI to do the same life they were already doing, just faster. They'll use it to write emails, summarize documents, generate images, and make presentations. They'll use it like a productive assistant. That's not what this is. That's like using a rocket engine to power a bicycle. You can do it, but you're missing the entire point of the engine.

"Here's the point. AI has moved from being a tool that does tasks to becoming infrastructure that produces intelligence. That sounds abstract until you picture it in the right way. Think about the last industrial revolution. You don't 'use electricity' the way you use an app. Electricity became the invisible layer under everything: factories, homes, transportation, communication. You didn't have to be an electrical engineer to benefit from it. But you also couldn't compete against a world that had it, pretending it was optional.

"Now in 2026 we're manufacturing something more valuable than electricity. We're manufacturing intelligence, not human intelligence, industrial intelligence – intelligence on tap, intelligence that can be scaled, copied, deployed, upgraded and embedded. As the cost of intelligence falls toward zero, the consequences are not incremental. They are explosive. Most people are not ready, because their minds are still trapped in an old economic equation: effort equals value. Time equals value. Credentials equal value. Scarcity equals value. In 2026, those equations get brutally renegotiated. If an AI system can produce in 10 seconds what used to take you 10 hours, the market doesn't regard your 10 hours anymore. It rewards the person who knows what to ask for, what to choose, what to reject, what to combine, what to deploy, and how to aim that power in the real world."

"AI becomes multimodal, in a way most people don't understand. Right now, many people think AI is 'text', or maybe 'text plus images'. But multimodal means something deeper. It means a machine that just doesn't read and write. It sees, hears, speaks, interprets, remembers context and acts across different forms of reality. It can take video, audio, text, sensor data, diagrams, blue prints, spreadsheets, and convert all of that into a single coherent understanding. You're not talking to a chatbot. You're talking to an intelligence that can parse the world the way you do, only faster, with more memory and with fewer emotional blind spots."

In the future, what if an AI system says it has consciousness and outputs spiritual insights? AI can never have a soul. The soul cannot be digitized. Souls cannot be created by software. **We are not bodies with souls. We are souls with bodies.** We aren't just biological machines. Christians will need to recognize a counterfeit "soul" if we see it. Know what is real and what is artificial.

Final Words: Reflections of an Aging Pastor

As I (Rev. Jerry) approach age 90, I personally struggle to understand the rapidly developing technology that engulfs me and perplexes me! Of course, my computer – savvy progeny regard my ignorance as the result of normal "cognitive slippage" for octogenarians. After all, our brains, as well as our bodies, experience shrinkage. Yet, in my heart, I catch a glimmer of this new technology and realize that I must try to understand the significance of AI. One man I met said, "'AI'? Why, I don't even know how to spell it." That speaks for me!

As a pastor, my first thought is to turn to the Bible. Of course, the Bible says nothing specific about AI because none of the Biblical writers knew anything about it. However, they knew about people. They knew about our human nature. Inspired by the Spirit of God, they wrote about that with wisdom, discernment, and prescience. And so I begin with a pivotal story told in the beginning of the Hebrew Bible. It's a story from the book of Genesis. You can profitably approach it either literally or metaphorically, and learn from it.

A Little Story from the Bible: Creation

"Then the Lord God formed a man [Adam] from the dust of the ground and breathed into his nostrils the breath of life, and the man became a living being. Now the Lord God had planted a garden in the east, in Eden; and there he put the man he had formed. The Lord God made all kinds of trees grow out of the ground – trees that were pleasing to the eye and good for food. In the middle of the garden were the tree of life and the tree of the knowledge of good and evil" (Genesis 2:7-9 NIV)

"The Lord God took the man and put him in the Garden of Eden to work it and take care of it. And the Lord God commanded the man, 'You are free to eat of any tree in the garden; but you must not eat from the tree of the knowledge of good and evil, for when you eat from it you will certainly die.'" (Genesis 2:115-17 NIV)

This was the Creator's command. He offered no explanation; He provided no rationale. He simply said, "This is the way I want you to live." Down through the echoing corridors of human time, the Eternal God is saying, "Choose my agenda or yours." In each generation we earthlings must make that choice. We are free to do things our way or His way. But it is important to understand that every choice we make has inevitable and irrevocable consequences. (i.e.. "for when you eat from it you will certainly die." (Genesis 2:17b NIV).

The "Fall" (From God's Grace)

"Now the serpent was more crafty than any of the wild animals the Lord God had made. He said to the woman [Eve], 'Did God really say, "You must not eat from any tree in the garden?"' The woman said to the serpent, 'We may eat fruit from the trees in the garden, but God did say, "You must not eat fruit from the tree that is in the middle of the garden, and you must not touch it, or you will die.' (Genesis 3:1-3 NIV)

Now notice how Evil (symbolized by a serpent) always, and deliberately, distorts the truth. He implied that a nasty Creator forbade the eating of any tree. Eve corrected him by saying that a providential God invited them to eat of every tree – except one. For on the day that they crossed that line they would surely die.

Evil (the serpent, Satan) countered, "'You will not certainly die,' the serpent said to the woman. 'For God knows that when you eat from it your eyes will be opened, and you will be like God, knowing good and evil.'" (Genesis 3:4-5 NIV)

This little story, told and re-told through the millennia, is crucial in understanding our relationships with one another and with God. We can ill afford to dismiss it as "just a little story". It's there in scripture for a purpose. It has often been quipped that, whereas we might read many books, the Bible is the only book that reads us! As I contemplate the "Temptation", I ponder the question, "Just what was the original Temptation?"

The "Temptation"

"When the woman saw that the fruit of the tree was good for food and pleasing to the eye, and also desirable for getting wisdom, she took some and ate it. She also gave some to her husband, who was with her, and he ate it." (Genesis 3:6 NIV)

I wondered, what was so tempting about this? What was this original temptation? What really tempted the couple? What was the "grabber"? I mean, after Eve was munching and crunching, what prompted Adam to reach for the forbidden fruit? Was it the thought of sharing a juicy piece of fruit or a juicy sidekick? Why did they disobey God's command? We all know now that nothing tastes as good as "forbidden fruit". But they might not have known that at this point. As I was reaching puberty, anything labelled "forbidden" was more enticing and sounded slightly naughty. But this initial temptation for Adam and Eve was far more sinister and menacing.

What was it? The color? The flavor? No. The temptation was that they could be like God and hold this power in their hands. The serpent said it, "For God knows that when you eat from it your eyes will be opened, and you will be like God, knowing good and evil." (Genesis 3:5 NIV) They wanted knowledge, and knowledge is power.

They wanted knowledge and knowledge is power! They could be like God, clasping his power! Not only that, they could – on their own – differentiate between good and evil. So, guess what? They didn't need God anymore. They could make their own decisions and figure out everything else for themselves. This is the primal temptation that has persisted in every generation.

So from my Christian perspective, I see the same temptation lurking in our confrontation with AI. We have the opportunity to reach out and pluck this magical power for ourselves. That's one thing. But we might follow that up by discarding God as irrelevant for our life. And that's another!

When Adam and Eve reached out to "grab all the gusto," they lost more than they gained. They lost their primordial paradise. They lost Eden. "After he drove the man out, he placed on the east side of the Garden of Eden cherubim and a flaming sword flashing back and forth to guard the way to the tree of life." (Genesis 3:24 NIV). They abandoned their Creator and His agenda. Just what was His agenda?

God's Agenda (through the witness of the prophets)

Through inspired prophets, spokesmen for God, He expressed His agenda. He said, in effect, that we are to learn to care for others, to be compassionate, to practice justice and righteousness, love and mercy, to be in fellowship with God. Here is just a sampling:

Isaiah 58:5-9a, NIV: "'Is this the kind of fast I have chosen, only a day for people to humble themselves? Is it only for bowing one's head like a reed and for lying in sackcloth and ashes? Is that what you call a fast, a day acceptable to the Lord?

"' Is not this the kind of fasting I have chosen: to loose the change of injustice and untie the chords of the yoke, to set the oppressed free and break every yoke? Is it not to share your food with the hungry and to provide the poor wanderer with shelter - when you see the naked, to clothe them, and not to turn away from your own flesh and blood?

"'Then your light will break forth like the dawn, and your healing will quickly appear; then your righteousness will go before you and the glory of the Lord will be your rear guard. Then you will call, and the Lord will answer; you will cry for help and he will say: here am I.'"

Jeremiah 31:31,33-34 NIV: "'The days are coming,' declares the Lord, when I will make a new covenant with the people of Israel and with the people of Judah. – – – This is the covenant I will make with the people of Israel – – – I will put my law in their minds and write it on their hearts. I will be their God, and they will be my people. No longer will they teach their neighbor, or say to one another, "Know the Lord," because they will all know me, from the least of them to the greatest – – –. For I will forgive their wickedness and will remember their sins no more.'"

Hosea 6:6 NIV: "For I desire mercy, not sacrifice, and acknowledgment of God rather than burnt offerings."

Hosea 12:6 NIV: "But you must return to your God; maintain love and justice, and wait for your God always."

Amos 5:21-24 NIV: "I hate, I despise your religious festivals; your assemblies are a stench to me. Even though you bring me burnt offerings and grain offerings, I will not accept them. – – – Away with the noise of your songs! I will not listen to the music of your harps. But let justice roll on like a river, righteousness like a never-failing stream!"

Micah 6:6-8 NIV: "With what shall I come before the Lord and bow down before the exalted God? Shall I come before him with burnt offerings, with calves a year old? Will the Lord be pleased with thousands of rams, with ten thousand rivers of olive oil? Shall I offer my firstborn for my transgressions, the fruit of my body for the sin of my soul? He has shown you, O mortal, what is good. And what does the Lord require of you? To act justly and to love mercy and to walk humbly with your God."

God's Agenda: Through the witness of Jesus

Christians advance the belief that "God was in Christ" (2 Corinthians 5:19a KJV). How God does that is a mystery. Many have expressed the thought that Jesus is "the human face of God," in other words, that in Jesus' words and actions we are shown the personality of God Himself. Trusting in this revelation becomes the key to restoration of fellowship with the God whose spirit Jesus embodies.

Hebrews 1:1-3a NIV: "In the past, God spoke to our ancestors through the prophets at many times and in various ways, but in these last days they had spoken to us by his Son, whom he appointed heir of all things, and through whom also he made the universe. The Son is the radiance of God's glory and the exact representation of his being, sustaining all things by his powerful word."

And what was God's agenda that was presented by this Jesus of Nazareth? He largely described it in "the Sermon on the Mount," told in Matthew, chapters 5, 6 and 7. It's most likely a compilation of the themes of many of Jesus' sermons preached many times in many places, but as a literary device it lumps them all together. Throughout my ministry, I have had people say to me, "I may not be 'religious' in the sense of belonging to an institutional church, but I consider myself 'spiritual'. I try to build my whole life on the

Sermon on the Mount."

I realize that these people expect me to be overwhelmed with admiration. Instead, I ask incredulously, "You build your whole life on the Sermon on the Mount?" "Yeah." "That's so wonderful. What does it say?" And they don't have a clue.

Then, in Matthew 25:31-40 Jesus talked about feeding the hungry and thirsty, welcoming the stranger, clothing the naked, visiting those sick and in prison. He said to us, between the lines, "This is God's agenda."

The Apostle Paul, following his Master, added these words in Romans 13:8-10 NIV: "Let no debt remain outstanding, except the continuing debt to love one another, for whoever loves others has fulfilled the law. The commandments, 'You shall not commit adultery,' 'You shall not murder,' 'You shall not steal,' 'You shall not covet,' and whatever other commands there may be, are summed up in this one command: 'Love your neighbor as yourself.' Love does no harm to a neighbor. Therefore, love is the fulfillment of the law."

Elsewhere Paul wrote, in Galatians 5:14, NIV: "For the entire law is fulfilled in keeping this one command: 'Love your neighbor as yourself.' "All of this is God's agenda for humankind. The first couple (historically or metaphorically) walks away from God, and we continue to abandon our Creator. We have perpetuated the sin: to have God's power while dispensing with God. When I consider all the blessings that we might derive from the use and application of AI, I am dazzled. But, as my co-author shrewdly perceives, we earthlings need biblical guidelines. Thousands of years ago the Psalmist wrote in Psalm 119:105: "Thy word is a lamp unto my feet, and a light unto my path."

Our Agenda

At an Easter Sunrise Service, 6:30 a.m., April 5, 2026, our astute Lay Leader, Mr. Kent Schellinger, Sr., assessed the naked vulnerability of our own human nature:

"Ever since the garden, humanity has been trying to cover what we cannot fix. As humanity grows in knowledge, we become remarkably powerful. We shape our environment. We cultivate the land. We harness energy. We build great cities, trade and create markets and institutions. We split atoms and map genomes. We solve problems our ancestors could not imagine.

"But there is something we do not solve. We shape our environment – but we cannot bend our will. We lengthen life – but we cannot conquer death. We can manage behavior – but we cannot purify our motives. We can write laws – but we cannot rewrite what is in our hearts. The law just reveals what is broken. With institutions we manage to restrain what is dangerous. With medicine we extend our lives. But none of them cure the deeper fracture first discovered in the garden." (Used with permission.)

Generally, you can count on the fact that we are out for ourselves. We are able to create tools to serve ourselves. Occasionally, we are able to weaponize them. There's an engrossing passage in Isaiah 14:13-15 NIV. Perhaps it originally referred to a Babylonian king, but – in a larger sense – it might represent Satan or our own selfish soul: "You said in your heart, 'I will ascend to the heavens; I will raise my throne above the stars of God; I will sit enthroned on the mount of assembly, on the utmost heights of Mount Zaphon. I will ascend above the tops of the clouds. I will make myself like the Most High.' But you are brought down to the realm of the dead, to the depths of the pit."

Now notice the string of "I wills"! That all too frequently reflects our own response to our world. So many of our efforts focus demonically on our own self-aggrandizement. As my coauthor, Dan, is careful to point out, AI is morally neutral. It by itself is neither good nor evil. Human beings, however, can be good or evil. Perhaps we can turn all this around.

One day, centuries after Isaiah 14 was written, Jesus would say to the Twelve: "I saw Satan fall like lightning from heaven." (Luke 10:18 NIV) He most likely was referring to the legend of Satan being cast out of heaven because of his pride and rebellion against God. Jesus' admonition to his disciples was a warning against human arrogance and audacity.

In Conclusion

From the Bible's viewpoint, we were created by God to have fellowship with Him and to serve Him. We were endowed with both intelligence and with freedom of choice. We created tools and technologies. We need to offer all of this back to the purposes of an all-wise God. Instead of declaring, "I am using AI to serve me," we ought to be affirming, "I am using AI to serve God." We are free to fulfill His eternal agenda for His creation.

In the 13th century, St. Francis of Assisi, Italy, wrote this beautiful prayer:

> "Lord, make me an instrument of Thy peace;
> Where there is hatred, let me sow love;
> Where there is injury, pardon;
> Where there is doubt, faith;
> Where there is despair, hope;
> Where there is darkness, light;
> And where there is sadness, joy.

"O Divine Master,
Grant that I may not so much seek
To be consoled, as to console;
To be understood, as to understand;
To be loved, as to love;
For it is in giving that we receive,
It is in pardoning that we are pardoned;
And it is in dying that we are born to eternal life"

It seems to me, an aging octogenarian, as well as a technologically-challenged pastor, that we might profitably take the first line of St. Frances' prayer, "Lord, make me an instrument of Thy peace," and extend it to prayerfully say, "Lord, make me, and my tools and technology, instruments of Thy peace." Amen

To Sum Up...

Scripture does not condemn tools or technology. God Himself commanded the building of the tabernacle with great skill and artistry. Solomon's temple stood as a marvel of engineering. Throughout history, God's people have used the best tools available to serve the love of God and neighbor. AI need not be different. Used rightly – with proper boundaries, genuine accountability, and humble recognition of its limits – it can serve these same ends.

But used wrongly, AI repeats humanity's oldest temptation: to trust our creations more than our Creator. The question before us is not whether we will use AI, but whether we will use it wisely, with eyes wide open to both its promises and its perils. The biblical witness provides clear guidance if we have the wisdom to heed it. In the end, the same God who warned against trusting in chariots calls us still to trust in Him alone, using whatever tools we have in submission to His greater purpose.

We have only scratched the surface here in this book, but it is our hope we will motivate you to delve deeper in artificial intelligence, how it is benefiting us today, and to be aware of potential misuses of this world-changing technology.

About The Authors

Dan Keen is a retired editor and publisher of a newspaper in Southern New Jersey. He has written many articles for national computer magazines, and books on a variety of subjects from computer programming to children's science for such publishers as Sterling Publishing, Tradewinds, and McGraw-Hill. Dan taught computer programming at Stockton State College in New Jersey. He is the former chief engineer for two Christian radio stations.

Rev. Jerry Crossley: I grew up in Philadelphia and was privileged to be part of a warm, loving fellowship that encouraged young people in their faith. I entered the ministry, eventually pastoring churches in the Pocono mountains, in the city, in small towns and villages. It has been the joy of my life. Whereas I could not even spell "science" until 6th grade, yet, as the years have passed by, I have gained many insights into the age-old truths of the Bible. In our book, Dan and I make the effort to bring both disciplines together. I write in the hope that an age-old story in the Bible will touch your heart and mind with a deeper truth, and that you will sense the drama between the Creator and His creation.

Other titles from Rev. Jerry Crossley

All You Ever Need is Jesus by Rev. Jerry C. Crossley

Full Gospel by Rev. Jerry C. Crossley

The Seen and the Unseen by Jerry Crossley

Journey to Jesus by Jerry Crossley

Shine Like Stars by Rev. Jerry C. Crossley

Holy Ways in Holidays by Rev. Jerry C. Crossley

Other titles from Higher Ground Books & Media:

Don't Be Stupid: And I Mean That in the Nicest Way by Rebecca Benston

It's Only a Game by Darrel Johnson

Raven Transcending Fear by Terri Kozlowski

The Power of Knowing by Jean Walters

What on Earth is Happening! by Terra Kern

One Step(pe) at a Time by Lori (Martin) Potts Zimmerman

Circling Around the Davidic Psalms by Stephen Shepherd

The Real Prison Diaries by Judy Frisby

Our Journey of Faith by Miranda Thornsberry

One Day in May by Joanne Piccari Coleman

Revealing His Might and Power by Deanna Rodriguez

Add these titles to your collection today!

http://www.highergroundbooksandmedia.com